Environmental Diplomacy

Saving the Sundarbans
and Restoring Indo-Bangladesh Friendship

Shloka Nath
Researcher, Gateway House
Indian Council on Global Relations

August 2011

Edited by Meera Kumar

GATEWAY HOUSE
INDIAN COUNCIL ON GLOBAL RELATIONS
भारतीय वैश्विक संबंध परिषद्

Published by
Gateway House: Indian Council on Global Relations
3rd floor Cecil Court, M.K. Bhushan Marg, Next to Regal Cinema, Colaba, Mumbai 400 039
T: +91 22 22023371   E: info@gatewayhouse.in
W: www.gatewayhouse.in

Gateway House: Indian Council on Global Relations is a foreign policy think tank in Mumbai, India, established to engage India's leading corporations and individuals in debate and scholarship on India's foreign policy and the nation's role in global affairs. Gateway House is independent, non-partisan and membership-based.

Printed by Spenta Multimedia, Mumbai
Front Cover Image: Fishing Boats in the Sundarbans
Source: Creative Commons / BriVos
Design & Layout: Divya Kottadiel

Contents

Acknowledgements

The author wishes to thank Jairam Ramesh (former Union Minister of Environment and Forests, India), Varad Pande (Officer of Special Duty to the Minister at Ministry of Environment and Forests, India), Ambassador Farooq Sobhan (former Foreign Secretary of Bangladesh), Geetanjoy Sahu (Assistant Professor at Tata Institute of Social Sciences), Ramaswamy R. Iyer (former Secretary Water Resources, Government of India), AK Raha (PCCF, WB Forest Department) Madhu Sarin (Forest Rights Activist), Professor Niaz Ahmed Khan (Country Representative, IUCN), and Dr. Haripriya Gundimeda (Associate Professor, Department of Humanities and Social Sciences, Indian Institute of Technology, Bombay) for sharing their valuable insight and support for this research paper. Special thanks to Manjeet Kripalani, (Founder and Executive Director, Gateway House), and Ambassador Neelam Deo (Founder and Director, Gateway House) and Meera Kumar for their unwavering support and encouragement, invaluable help in commenting on my drafts, suggesting new directions for research, and pointing me to additional sources. This research would not have been possible without Nehal Sanghavi, Samyukta Lakshman, Celine D'Silva, Madhura Joshi and the entire team at Gateway House whose support and motivation remains unmatched.

About the Author

Shloka Nath is a Researcher in Gateway House's Studies Programme. Prior to this she was Principal Correspondent with Forbes India, specialising in Strategic Affairs, Financial Inclusion and Business and Law. A graduate from the London School of Economics and Political Science with a BSc in Government, she has worked as a broadcast journalist with the BBC in London and as an Anchor and News Correspondent with New Delhi Television (NDTV) in Mumbai. She was a speechwriter at the House of Lords and during her tenure, successfully established an All Party Parliamentary Group for Entrepreneurs. Shloka has also worked on promoting press freedom worldwide for the Committee to Protect Journalists (CPJ) in New York.

Shloka is currently studying Public Policy at Harvard's Kennedy School of Government.

Executive Summary

In January 2010, India's then Minister of Environment suggested that India and Bangladesh join hands to protect the Sundarbans from environmental degradation. The proposed Indo-Bangladesh Sundarbans Eco-System Forum, which is currently in the planning stages, is to be made functional later this year. The forum, which will include non-governmental organisations and civil society of both the countries, plans to coordinate efforts in afforestation, management of mangroves and conservation of the tiger. It is an innovative idea and the first of its kind for India in using bilateral environmental problems to foster broader regional cooperation between India and Bangladesh.

The Sundarbans sits on the sensitive border between India and Bangladesh. It is the world's largest mangrove forest but also one of the most endangered eco-systems in the world. Many of the problems that have plagued relations between India and Bangladesh over the last four decades lie in the Sundarbans. Resolving these concerns will help advance their relationship and save a vitally important global ecosystem.

Gateway House: Indian Council on Global Relations proposes a bold new plan to accomplish this by taking the ministry's efforts a step further. While we support the joint Indo-Bangla forum, we believe that is not sufficient to save the Sundarbans and repair Indo-Bangla relations. Instead, we propose creating the forum through an Indo-Bangla Bilateral Environmental Treaty for the Sundarbans. A bilateral treaty will facilitate implementation of the programmes under the forum. More importantly, it will not restrict itself to the agenda of the forum alone. Rather, it will make space for more inclusive and coordinated reform between the two nations across state, district and grassroots levels.

For instance, last year the Union Cabinet of India approved the Rs. 1,156 crore Integrated Coastal Zone Management project. Of this, Rs. 300 crore will be spent in West Bengal, most of it on the Indian Sunderbans. The project, to be funded by the World Bank, will be executed over five years and includes prevention of erosion of the islands, building of storm shelters, promotion of eco-tourism and improving the livelihood of the inhabitants of the region. In addition, the 13th Finance Commission has also sanctioned a grant of Rs. 450 crore in 2010, for strengthening embankments at critical areas in the Indian Sunderbans. We at Gateway House believe it would be more efficient to bring these proposals under a single, cohesive banner, to be executed by both India and Bangladesh simultaneously, in order to have maximum impact. Which is why a cross-border initiative such as an Indo-Bangla Treaty is so necessary; it will take into account the sensitive ecosystem as a whole, rather than piecemeal efforts with both India and Bangladesh working in silos, as is currently the case.

In this paper we first examine the various causes of disagreement between India and Bangladesh, then make recommendations for their resolution.

The most contentious problem in Indo-Bangla relations has been the construction of the Farakka Barrage in 1975 and the allocation of sharing of the Ganges water. This diverted the Ganges waters upstream, adversely impacting the Sundarbans. Most parts of the wetlands have now surpassed their water-salinity thresholds and degraded much of the fragile ecosystem.

Over the next few decades, a matter for grave concern for both nations will be the threat of "climate refugees" moving across the border into India from Bangladesh, as well as internal migration from areas like the Sundarbans to the slums of cities like Kolkata and Dhaka. Bangladesh, a low-lying deltaic region and one of the most densely populated countries in the world faces danger of being partially submerged if there is multi-foot rise in sea level as a result of climate change. This could result in between 10 to 30 million individuals displaced along its southern coast, turning them into "climate refugees." The Sundarbans, where continued destruction due to rising sea levels and natural disasters has already produced thousands of climate refugees, is at the epicentre of this pitched battle against climate change. For all its predicaments, the Sundarbans is a place where adapting to climate change actually seems possible, thanks largely in part due to the one commodity that both India and Bangladesh have in abundance; human resilience. If India and Bangladesh were to work together to save this fragile region, rather than pitying the Sundarbans, the world can learn from its example. What is needed to resolve the problems is a new understanding that will encompass the economic and the environmental. It will mandate effective cross border management in both countries at two levels: state and local. Community-driven projects aimed at reducing unsustainable livelihood practices that are causing environmental degradation are an imperative. For the treaty to be effective, there must be state level implementation of policies to ensure joint cooperation in addressing and monitoring of problems. For instance, India and Bangladesh can begin a joint relocation and emergency evacuation programme for climate disasters such as cyclones or flooding. This will go a long way in preventing climate refugees.

Both nations can use their strong grassroots institutions to ensure policies are practically and effectively implemented. Working together under a bilateral agreement, India and Bangladesh can potentially manage the crises of climate change refugees through the use of micro-credit programs like micro-loans for livelihoods and micro-insurance for environmental disaster cover. Sharing of the water in the Ganges–Brahmaputra basin is a historic opportunity for both countries to create not only a bilateral, but also a regional policy, under the auspices of a body such as the South Asian Association for Regional Cooperation (SAARC). The focus on an Environmental Impact Assessment (EIA) for the entire Ganges-Brahmaputra Delta will be a good starting point.

This paper uses examples from around the world of successful trans-boundary environmental agreements. The International Gorilla Conservation Programme (IGCP) between Uganda, Democratic Republic of Congo (DRC) and Rwanda is an excellent illustration of cross-border management of forest reserves. The Border Environment Cooperation Commission (BECC) between U.S.A and Mexico is unique in that it is largely driven on community participation. Our paper suggests, in addition, that the state government of West Bengal must be directly involved in the management of the Sundarbans under a bilateral agreement. The state can play a key role in scientific research on mangrove protection and biosphere management, funds for which should be allocated under the bilateral agreement to set up a national institute for mangroves and coastal protection.

In September this year, the Indian Prime Minister, Manmohan Singh is set to visit Bangladesh for a highly anticipated round of talks. In a strategic move, West Bengal Chief Minister, Mamata Banerjee, will accompany him. Both sides are expected to sign a number of bilateral pacts, including a 15-year interim water sharing agreement over the Teesta River.

We hope that this paper and the recommendations contained herein will both improve bilateral relations between two important South Asian neighbours and lead to intelligent management and protection of the Sundarbans.

Introduction

In January 2010, Jairam Ramesh, then India's Environment Minister, while on a visit to Kolkata, displayed once again his penchant for the big idea. Together with his counterpart from Bangladesh, he announced that both countries were joining hands to protect the Sundarbans from environmental degradation and stated that the proposed Indo-Bangladesh Sundarbans Eco-System Forum, which is currently in the planning stages, will be made functional by the second half of 2011. In October 2010, a draft protocol to be signed by both nations was approved by the Bangladesh cabinet at a meeting chaired by Prime Minister Sheikh Hasina. It is an innovative idea and the first of its kind in using bilateral environmental concerns to foster broader regional cooperation between India and Bangladesh. For India and Bangladesh, this is a unique foreign policy opportunity.

In an exclusive interview with Gateway House in September 2010, Ramesh said, "Environmental diplomacy and environmental cooperation can often be triggers for enhancing broader regional cooperation. It helps to build trust, gets your people working with each other, learning from each other, and breaks down barriers. I hope to see more of this going forward."

The timing of the statement could not have been better; if the aspiration of Jairam Ramesh can be realized, not only will the fractious Indo-Bangla relationship be repaired, but also, the Sundarbans, the world's largest mangrove forest spanning a vast territory of over 10,000 sq kms, can be saved.

Sitting on the sensitive border between India and Bangladesh, the Sundarbans is one of the most endangered eco-systems in the world. Approximately 60 percent lies in Bangladesh and 40 percent in India. The problems that surround it have the potential to advance the relationship between these two important South Asian neighbours; given the magnitude of difficulties, the converse is equally true.

The region has long been a diplomatic challenge for both nations. Take for instance, the 30-year battle for control over

New Moore Island and sporadic disputes over freshwater diversions such as the Farakka Barrage in the Ganges Delta region, that many say have cut off water supply to Bangladesh and degraded the Sundarbans. Or the massive construction project that India has undertaken to fence itself off from its neighbour, along India's 2,544-mile border with Bangladesh, for which India has received much criticism.

But none of this will eventually matter if the rising water levels in the region engulf the delta, resulting in submergence of villages, "climate change refugees", and the loss of a unique eco-system and natural resources vital to two countries with such large populations. In September 2011, Manmohan Singh, the Indian Prime Minister will visit Bangladesh in an attempt to inject fresh dynamism into the relationship between the two neighbours. Both sides are likely to sign several bilateral pacts including the 15-year interim water sharing agreement over the Teesta River. Accompanying the Prime Minister to Dhaka will be the newly elected Chief Minister of West Bengal, Mamata Banerjee.[i] This is the ideal opportunity for both sides to also take a fresh look at the problems concerning the Sundarbans delta and to bring into effect an environmental bilateral agreement to manage the fragile ecosystem.

Fortunately, there are numerous international examples of trans-border environmental agreements that will serve as good models for India and Bangladesh. Finland and Russia have been at the forefront of environmental cooperation for many years. With concerns over oil spills, maritime transport and biodiversity degradation, both countries regularly exchange information and technology and conduct seminars on implementation of the Kyoto protocol. There are other examples as well which are discussed in the Conclusions section. In this paper we examine how India and Bangladesh could work together to restore good relations and save the second largest delta in the world. The Sundarbans require a cross-border, multi-layered approach to addressing its problems, at both: the local and state level. What is needed is effective cross- border management through community driven projects aimed at reducing unsustainable livelihood practices that are causing environmental degradation. What is also necessary is successful state level implementation of policies that ensure joint cooperation in addressing and monitoring the problems faced by the people of this region. Crucially, there must be cross-border coordination between various district administrations in areas along the porous Sundarbans border such as Khulna and Satkhira districts in Bangladesh and Murshidabad and North 24 Parganas in India, to name a few.

An Analysis of the Problems

Background

The relationship between India and Bangladesh has been tainted by differences over control of islands in the delta, the maritime boundary, and brewing disagreements over water-sharing rights. Bangladesh has refused to sell India some of its gas reserves; an India-Myanmar pipeline through Bangladesh has not materialized and it takes more than a week for a truck-load of goods to cross the border in either direction. India had proposed a free trade agreement to help soften the trade imbalance, which Bangladesh rejected. Since the mid 1980s Bangladeshi presidents, upon election, have made Beijing, not India their first port of call, thereby strengthening Sino-Bangla relationship to a level where trade between the two countries has risen from $715 million in 2000 to $5 billion in 2010; in 2009 China replaced India as Bangladesh's largest trading partner.[ii]

Sheikh Hasina, elected as Prime Minister of Bangladesh in 2008, has made a notable effort to improve India-Bangladesh relations, from the bitter animosity that prevailed during Begum Khaleda Zia's second term as Prime Minister from 2001 to 2006. The agreements signed during Sheikh Hasina's recent visit to India ranged from cooperation in combating terrorism, removal of tariff and non-tariff barriers to trade, and fair demarcation of maritime boundaries between the two countries, to a more humanitarian and sympathetic border management. New Delhi responded warmly, with a decision to sell 250 MW of power to Bangladesh and extend a line of credit of US $1 billion.[iii] But since then, Sheikh Hasina has also overseen a crackdown on Indian separatist groups from the northeast who have traditionally found shelter in Bangladesh – a major sticking point in relations.

Unfortunately the Sundarbans issue was pushed to the rear with little or no discussion around it. Ramesh's robust, lateral environmental diplomacy, therefore, is welcome.

Climate Refugees

Nearly 80 percent of Bangladesh's densely populated mainland is at sea level, pitting its citizens at the frontlines of the global battle against climate change. [iv] The immediate concern is that of "climate refugees." While the term "climate refugees" has, as yet, no place in international law,[v] according to the UN's Intergovernmental Panel on Climate Change, there are around 25 million climate refugees, which can increase to as many as 150 million by 2050.[vi] According to the International Organization for Migration, these individuals are '...almost invisible in the international system...unable to prove political persecution in their country of origin they fall through the cracks of asylum law'.[vii] A major concern for India and Bangladesh lies in determining responsibility for the protection and rehabilitation of these refugees, most of who currently reside in slums in Dhaka or Kolkata.

A resolution can begin in the international arena through an organization like the United Nations if India and Bangladesh rally the international community to acknowledge their status under law, thereby guaranteeing that individuals displaced by climate change are entitled to natural rights under an international treaty. Bangladesh has taken several steps forward, clamouring for the international community's attention. In December 2008 member states of the UN Climate Change Convention held their 14th summit in the Polish city of Poznan. Bangladesh insisted that the prospect of international migration for climate change victims should be incorporated in the new global climate deal being deliberated. [viii] Also necessary is a joint relocation and emergency evacuation program in the case of climate disasters such as cyclones or floods - both of which are becoming annual events.

Both India and Bangladesh's responses to the onslaught of climate disasters has been mostly relief-centric and reactive in the past, littered with examples of ineffectual implementation of national and state level policies. India has created national level policies such as the Disaster Management Act in 2005 to deal with climate change disasters. The Cabinet Committee on Management of Natural calamities and The National Disaster Response force were constituted the same year. But these task forces have had limited effect in curtailing the devastating effects of climate disasters as evinced by the 2008 Kosi floods. Most notably, the National Action Plan for Climate Change makes no mention of any plan to save the biodiversity of the Sundarbans.[ix] In March 2009, the World Wildlife Fund, India, set up a Climate Adaptation Centre in the Sundarbans in Mousini Island with an Electronic Early Warning system to alert villagers to oncoming cyclones. The objective was to train the inhabitants in climate adaptation strategies and to set up requisite infrastructure to shield them from disasters. However, Cyclone Aila, which hit the fragile deltaic region in May the same year, destroyed all the WWF's work in Mousini.[x]

Last year the Ministry of Environment approved a $200-million World Bank-supported Integrated Coastal Zone Management (ICZM) Program that will look into building capacity of coastal areas including the Sundarbans and will be completed by 2015. However, while implementation of the ICZM project in the states of Gujarat and Orissa has begun reasonably, the World Bank reports less than satisfactory progress in West Bengal.[xi] Bangladesh has also had limited success in disaster management. Cyclone Aila caused extensive devastation in the Delta country and exposed some of the government's weaknesses in responding to disaster situations effectively; namely, the coordination of local and state level agencies. In July this year, The Inter-ministerial Disaster Management Coordination Committee met to approve the draft version of the Disaster Management Act, a bill that will go a long way in securing efficient and timely state responses to disasters as well as train local communities in disaster management strategies and protection. The Bill thus represents a paradigm shift from a post disaster response mechanism to that of ensuring early preparation for risk reduction.[xii]

Environmental Impact

The Sundarbans is the most biologically productive of all natural eco systems, and is home to a stunning variety of fauna, many threatened reptiles and as of 2003, more than 600 Royal Bengal tigers, the last few remaining in the world.

Speaking exclusively to Gateway House, former Bangladesh Ambassador to the UN, Harun Rashid says, "A framework agreement is necessary between the governments of Bangladesh and India to establish institutional linkages to facilitate the sharing of knowledge, information and capacity building programs in preserving the flora and fauna in the Sundarbans including its wetlands, mangroves and biosphere protection and management." In his view, a joint committee of climate and biodiversity experts from Bangladesh and India should harness the knowledge of local communities at the grassroots level on how they deal with the ongoing changes in climate and grow crops. "In some coastal areas of Bangladesh local farmers have adopted innovative methods to grow fruits and vegetables in inter-tidal areas and such knowledge may help in saving the Sundarbans," says Rashid.

The Sundarbans are a major source of timber and natural resources which protect the land from frequent storms and the most important source of the fish and shrimp on India's eastern coast. In Bangladesh the shrimp industry accounts for 5% of GDP, over $ 300M in exports and 1.2 million jobs.[xiii] But pollution from industry, oil spills and a potential dispute over exploitation of timber resources and agricultural encroachment on the eastern and western boundaries of the Sundarbans are a huge threat to the fragile ecosystem.

Smuggling between India and Bangladesh is a common phenomenon. In 2000, a Central Law Commission Report in India said that illegal trade between India and Bangladesh was approximately US $5 billion.[xiv] Unfortunately, state-led initiatives to combat these problems have not always been successful given that the entry of most migrants is clandestine; they have linguistic, ethnic, and religious similarities with the local populations. State governments admit their resources are limited and the challenges are numerous. In July 2011, incidents of child smuggling were reported as poverty stricken young boys posing as bootleggers cross the porous Indo-Bangla border.[xv] According to these reports, locals also allege that border security guards work in tandem with smuggling cartels.[xvi] Recently, the West Bengal Crime Investigation Department (CID) released a report assessing the role of Bangladeshi money behind the illegal tiger hunting trade in the forests of the Indian Sundarbans. According to the study, Bangladeshi moneylenders are financing poachers, arming them with local boats and firearms to hunt tigers in the Sundarbans, most of the time disguised as Indian fishermen.[xvii]

Unemployment

The total population of the Sundarbans is approximately 4.5 million in India and over 6 million in and around the forests in Bangladesh.[xviii][xix] According to the 2001 census, a majority of the total population in the Indian Sundarbans remains unemployed and severely impoverished. Most are heavily dependent on the forest, eking out a living by selling materials extracted from the forest. [xx] In Bangladesh, one study estimates that 3.5 million people surrounding the area are directly or indirectly dependent on the Sundarbans ecosystem.[xxi] "The pressure on the resource utilization can be eased to the extent that alternative livelihoods are available. This is a challenge for both government bodies and NGOs providing education, microcredit and access to market," says Md. Tamimul Alam Chowdhury, in a report for the Centre for River Basin Organizations and Management, Indonesia.[xxii]

A robust program in this regard has been undertaken by Bangladesh, and can be replicated by India. Bangladesh's Nishorgo program of the Forest Department (FD) was launched in 2004, with support from USAID and the European Commission, among others. Nishorgo's goal is to have the forest department co-manage the administration of the resources of the Sundarbans with key local, regional and national stakeholders from the Wildlife Advisory Board, other ministry officials, NGOs, journalists, and even teachers.[xxiii]

But Nishorgo has its problems. While co-management between different stakeholders is a good concept, it has to be done at the grassroots level rather than through a top-down approach. Eventually, coordination with similar efforts from India will produce the best results.

Less than two hundred years ago, the Sundarbans extended all the way from the Kolkata Bay to the east and from Burma to the west, almost till Dhaka in the north. Today the dense forests no longer dominate the delta; there are empty spots where trees once stood. With immediate collaboration and strict implementation of recommendations between India and Bangladesh, the further retreat of the Sundarbans can be halted. Instead of being on the diplomatic backburner, with sufficient political will this could spur bilateral cooperation between the two countries.

CHAPTER ONE

History and Determinants of Indo-Bangladesh Foreign Policy: A Fractured Relationship

Bangladesh suffers from an identity crisis stemming from different definitions of nationalism, and the struggle for supremacy between the Bengali identity and Bangladeshi nationalism.[xxiv] This identity crisis has also defined to a great extent its foreign policy with regard to India. The fractured relationship between the two neighbours has resulted in inertia over bilateral problems like the Sundarbans.

This was well exemplified in 2011, when a film, *Meherjaan*, premiered in Dhaka. Directed by a young Bangladeshi woman, the film showcases actors from India, Pakistan and Bangladesh and depicts the lives of three different women during the 1971 war. But viewers expressed anger that the film had inverted the traditional war narrative; it was the story of a good Pakistani soldier and a bad Liberation fighter. Predictably, the film was withdrawn from theatres a few weeks later. [xxv]

This incident is telling. Bangladesh's attempt to interpret the cruelties of its bloody birth in 1971 is really an identity crisis expressed largely in the narratives of nationalist glory.[xxvi]

Determinant 1: History

In 1971 Indira Gandhi sent Indian troops to fight the Pakistan army in support of the Bangladesh war of independence. The intervention concluded the war in nine short days, ending a nine-month campaign of genocide and ethnic cleansing that had left countless dead and many millions displaced. Immediately after his return from Pakistan in January 1972, Mujibur Rahman declared that Bangladesh was to have special ties with India and openly endorsed the principles of non-alignment, peaceful co-existence, and opposition to colonialism. The love affair between India and Bangladesh seemed destined to last forever. [xxviixxviii]

But as with all great romances, the relationship soon turned sour. Having swooped in to battle on behalf of Bangladesh, the Indian army did what all armies do; they behaved like triumphant soldiers. The treaty signed on December 16, 1971 was between an Indian general and a Pakistani general. Unexpectedly, Bangladesh's freedom fighters were no more than a postscript in what had turned out to be yet another scuffle between the two elder children of partition. [xxix] Bangladeshi troops were then forced to hand over their weapons to the Indian army. It would prove to be a wound that would fester for years to come.

On the other hand, India has felt a deep sense of betrayal that Bangladesh proved ungrateful. India had after all, hosted ten million Bangladeshi refugees; around three thousand Indian soldiers had given their lives to help liberate Bangladesh, and India had risked war with one of the most powerful countries in the world, the United States. [xxx] The treaty was signed between Indian and Pakistani generals, as there was no Bangladesh Army in existence at the time and leaving the freedom fighters armed could have led to a military coup in the first year of Bangladesh independence.

The determinants in Indo-Bangladesh foreign policy have antecedents prior to the civil war in 1971. Bangladeshis perceive their struggle as having culminated in 1971, a struggle for Bengali nationalism, separate and distinct from the Hindu majority and other religions in India, which began in the mid-1930s. [xxxi]

This struggle and subsequent anti-India sentiment was enunciated than in the 1940 Lahore Resolution of the All India Muslim League. Presented by A.K. Fuzlul Haq, it called for rights for Bengali Muslims in Bengal and in the creation of two Muslim states. But these expectations were later stifled when Mohammad Ali Jinnah declared Urdu as the state language in January 1948. The language movement would soon resuscitate the cause of Bengali nationalism, when on 21 February 1952, students of Dhaka University agitating for their language, faced police firing. Adding to growing unrest in the region was the decision by the Pakistan government to ban the poems of Rabindranath Tagore. [xxxii]

Post independence these legacies firmly attached themselves to Indo-Bangladesh relations. However, in July this year, in an attempt to re-establish historical links, UPA Chairman and president of the Indian National Congress, Sonia Gandhi visited Dhaka. In a high profile ceremony attended by

Bangladesh President Zillur Rahman, she received the Bangladesh Swadhinata Sammanona, or the Bangladesh Freedom Award, on behalf of late Prime Minister Indira Gandhi, for her tremendous contribution to the Bangladesh Liberation War in 1971. Prime Minister Sheikh Hasina described Indira Gandhi as "a true and great friend" and paid homage to the Indian soldiers martyred in the war. It is a sign of a promising future between the two neighbours, looking beyond a history of mistrust and colonial shadows.[xxxiii]

Determinant 2: Geography

India surrounds Bangladesh on three-sides in what has been called, "a great bear-hug of a border."[xxxiv]Bangladesh thus shares more than 90 percent of its international border with India. West Bengal covers the Western portion of Bangladesh, Assam and Meghalaya lie to the North and Tripura and Mizoram are on the East. This looming presence of a hefty neighbour has charted the course of both political as well as social considerations between the two nations. Nowhere is this more apparent than in the disagreements over water sharing (Chapter 2), dispute over maritime boundaries (Chapter 3) and the tenuous security and trade arrangements along with illegal immigration (Chapter 4).

A pertinent example would be that of the treatment of tribal peoples in the Chittagong Hill Tracts (CHTs) in Bangladesh. Viewed by the Bangladeshi government as pro-India, these communities have existed in the same areas prior to partition, but they are considered undesirable in the Bengali Muslim mainland. Subsequent settlement policies adopted by Gen. Ziaur Rahman and Gen. H.M. Ershad ended up marginalizing the tribal people who eventually fled to India to escape persecution. India then encouraged these migrants such as the Shanti Bahini, and Chakma refugees, to fight back. But once they were rehabilitated in the CHT, there was a knock-on effect of insurgency in the North Eastern territories of India. [xxxv] "The root cause of the crisis was the hunger for land in Bangladesh where the Bengali Muslims were (and are) in a majority but ethnic and religious factors added emotional and psychological intensity to the controversy," says Former Foreign Secretary J.N. Dixit.[xxxvi]

Determinant 3: Religion

Religious fervour and sentiment has also played a crucial role in crafting foreign policy for the two nations. In the 1990s both nations experienced a rise in religious fundamentalist parties. The demolition of the Babri Masjid in 1992 was a turning point in relations. Bangladeshis were aghast at the riots that followed; it portended that India would never accept a truly independent Bangladesh.[xxxvii]

India's frustrations with Bangladesh have manifested themselves with regard to the rise of Islamic extremism in the country, specifically the involvement of the ISI of Pakistan with India as the objective. [xxxviii]

The Hindu community became the target of widespread violence after the 2001 elections for having supported the Awami League and since 1971; the Hindu population in Bangladesh has been steadily dwindling as a result of sustained discrimination by the state. [xxxix] Confounding matters for India has been the Bangladeshi government's unwillingness to engage in productive discussions to tackle the problem.[xl]

Determinant 4: Leadership

Mujib adopted the policies of democracy, nationalism and secularism whereas Maulana Bhasani maintained an anti India stance in national policies, criticizing Mujib for signing the Friendship Treaty with Indira Gandhi. His brand of "Islamic Socialism" asserted that 14 percent of the Hindus were exploiting 86 percent of the Muslims of Bangladesh and that *Bangassam*, the amalgamation of Bengali speaking parts of East India and Bangladesh into a new state would render India weaker, a concept that still prevails in some parts of Bangladesh.[xli] In India on the other hand, there is a sense that Prime Minister I.K. Gujral gave away too much in the face of Bangladeshi obduracy.

In recent times Sheikh Hasina and her government appear to have concluded that if Bangladesh has to develop as an independent secular democracy, there is need for a South Asian model of governance; one in which India can be a partner of choice.[xlii]

CHAPTER TWO

The Ganga Water Disagreement

"That a region so richly endowed (with water) should remain so poorly developed is a painful paradox. The logic of optimum development and management of vast natural resources for national and regional benefit has been obscured by political boundaries, perceptional differences and a legacy of mistrust."
    - R. Rangachari and B.G Verghese [xliii]

The Ganges-Brahmaputra Delta and its river systems is the second largest hydrologic region in the world. The basin is approximately 1.75 million km^2 and spans five nations: Bangladesh, Bhutan, India, Nepal and China. It has a population of over 600 million people.[xliv] The Delta basin is an incredibly backward and poverty-ridden region; access to sanitation is scarce for the 250 million who live on less than US $2 per day[xlv] and social indicators, such as life expectancy and infant mortality are much lower than the world's average. [xlvi] According to the UN, it has the largest percentage of people living in poverty; nearly half of the 535 million who populate the basin are poor.[xlvii] However, the region is very rich in water; although the water, abundant during the Southwest Monsoon from June to October, is inadequate during the rest of the year. About 3,500 m^3 of water are available per capita annually in the GBM Basin. [xlviii] A paradoxical situation persists as this excess water also causes seasonal and devastating floods in Bangladesh. [xlix] One-third of the country floods annually during the monsoon season.[l]

One of the most persistent foreign policy disputes between India and Bangladesh has been the differences over water sharing of the rivers in the Ganges-Brahmaputra Delta. In 1975 India constructed the Farakka Barrage across the Ganges River, diverting water in order to replenish a parched Calcutta Port. As a result of the upstream diversion, the impact on the Sundarbans has been considerable. According to some estimates, the impact of the construction of the Farakka Barrage on dry seasonal flows and water salinity levels in the Sundarbans has been comparable to and potentially even higher than the effect calculated to occur from climate change a few decades later.[li] According to a UNESCO report this barrage diversion induced a decrease of 40% of the dry season flow.[lii][liii] Also exacerbating the situation is that the countries of the Himalayan River Basin through which both the Ganga and

Bramhaputra rivers flow, will face huge depletion in water availability on per capita cubic metre basis in the next few decades. According to a recent report, the per capita cubic metre basis is likely to decline from 7,320 to 5,700 in case of Bangladesh in 2030 and from 1,730 to 1,240 in case of India.[liv] The Ganges River arrives in Bangladesh after approximately 17 km down the Farakka Dam; even before reaching the Farakka, a certain percentage of the Ganges water is diverted for use in India, for irrigation by the Upper and Lower Ganges canals, pumped canals in Dalman, Bhapali and Zamania as well as numerous other, much smaller, withdrawals.[lv]

A recent study by Islam and Gnauck revealed astonishing results; water salinity levels in the Sundarbans had increased significantly in 1976 (post the construction of the Barrage) compared to that in 1968 before the barrage was constructed. [lvi] The soil and river water salinity data also shows most parts of the Sundarbans wetlands have surpassed the water salinity thresholds. In order to keep salinity intrusion at a level where the impact to the Sundarbans is negligible, water flow to the Sundarbans must be above 1,500 m^3/s. In 1975, it was found that the reduced water flow line and increased salinity line crossed each other at this optimum point.[lvii]

High salinity levels in parts of the Sundarbans have degraded much of the fragile ecosystem. Increased alkalinity has changed the structure and texture of the soil, leaving it infertile, with destroyed surface organic matter. A decreased flow of water from upstream sources has also severely impacted coastal food security. The same study asserts that by building the Farakka Barrage and diverting water for India's needs, there was a likely loss in rice output of 236,000 metric tons in 1976. [lviii] Salinity has also impacted and doubtless reduced the prospects for additional irrigation and fresh groundwater for both industrial as well as individual consumption needs. Water wells in coastal areas, such as the Sundarbans must penetrate more than 250 metres at a minimum to reach drinking water.[lix] The Bangladeshi section of the India-Bangladesh Joint River Commission has calculated that the consolidated financial losses of Bangladesh due to Farakka withdrawal from 1976 to 1993 amount to 113,240 million taka (nearly US$ 3 billion) in accordance with the 1991 price index.[lx] The effects of the Farakka Barrage however, are not limited to Bangladesh alone. Upstream and downstream districts in India have suffered the devastating effects of high sedimentation, increased flood intensity and collapsing embankments. In two districts in West Bengal - Malda and Murshidabad - the inhabitants have suffered large-scale population displacement and loss of livelihood. [lxi] Murshidabad, for instance, has a per capita water availability of a mere 525.63 cubic metres in 2009, an indication of water scarcity. [lxii] In Bihar, nearly 98 percent of the fisheries associated with the Ganges River and connecting wetlands, collapsed after construction of the Farakka Barrage.[lxiii] As stated in the report, "Floods usually get more highlighted in the national and international arena but one has to remember that its consequences are of short range as economic recovery is possible within a predictable time. In this connection the opinion of the riparians is that floods bring fertile soil helping in ripe harvesting. But the slow and steady disaster of riverbank erosion has a permanent effect upon the socio-economic conditions and demographic dislocation." [lxiv]

Pollution offloads from Bangladesh as well as India, such as industrial effluents, agrochemicals and domestic waste, result in trans-boundary water quality problems

Pollution offloads from Bangladesh as well as India, such as industrial effluents, agrochemicals and domestic waste, result in trans-boundary water quality problems. These are increasing to dangerous levels especially in the drier seasons when water flow is lower. A 2004 study has pointed out that these levels are both very harmful and totally unacceptable.[lxv] This poor surface quality of water, is leading to more removal of groundwater for both agricultural and human consumption. As a result, Bangladesh has faced dangerously high levels of mineral arsenic in its groundwater in areas of the Delta with alluvial soil. The problem was first identified in the 1980s.[lxvi]

Reduction in the downstream discharge has led to top-dying diseases amongst the fauna in the Sundarbans. A shortage of water has increased the siltation in the Sundarbans and the dense roots and the pneumatophores of the mangroves have also abetted this trapping of sediment. [lxvii] Degradation of the mangrove wetlands has also resulted in endangering species such as the Hog Deer, as well as their extinction. The Sundarbans has lost several species in the last century.[lxviii]

History of the Dispute

"Rivers have a perverse habit of wandering across borders . . and nation states have a perverse habit of treating whatever portion of them flows within their borders as a national resource at their sovereign disposal."
- John Waterbury, Hydropolitics of the Nile Valley (1978) [lxix]

On December 12, 1996, India and Bangladesh signed a historic treaty. The Ganges Water Treaty, valid for 30 years, established a framework for both nations to share the waters of the Ganges River. However, reactions to the treaty in both countries were fierce. [lxx] According to reports in India, members of the Bharatiya Janata Party (BJP), the Congress, and the Communist Party of India –Marxist (CPI- M) were unhappy with different aspects of the treaty. The Central government was accused of forfeiting West Bengal's interests; the BJP was worried that by stipulating the quantity of water to be shared over a percentage-based quota might result in problems for India during the dry season.[lxxi] Indian media flinched at the treaty, carrying several reports that the data on which the treaty was based was outdated and that it was biased in favour of Bangladesh.[lxxii]

Similarly, in Bangladesh the main opposition party, the Bangladesh Nationalist Party (BNP) was equally unhappy and criticized the government for tilting in India's favour. [lxxiii] Some press reports also alleged massive treaty violations by India, and in 1997 there were accusations against West Bengal for diverting more than its fair share of water to the Kolkata port.[lxxiv]

The problems with the dam began in 1950 when officials in India first began mulling the idea of building a dam at Farakka. And in 1962 India finally gathered the necessary financial resources to begin construction of the Farakka Barrage. It took India eight years and US $208 million to complete the project.[lxxv] Over the course of the next decade, India and Pakistan met several times to exchange data and discuss matters of technical expertise. The 1970 meeting is significant as it was the first time that India acknowledged the Ganges as an international river, thereby accepting that its water had to be shared. "Thus it took almost 20 years for an upstream nation to shift from the notion of territorial sovereignty to gradually accommodate restricted sovereignty," says Rakesh Tiwary, in *Conflicts over International Waters*, "Economic and Political Weekly." [lxxvi]

Many experts have defined the two decades from 1950 to 1971 as being futile in producing any agreements over sharing the water of the Ganges River. The reasons for this are manifold; some experts assert that the Pakistan government avoided the Farakka problem deliberately as it was more interested in working with India in negotiations over the Indus River basin. Others, such as B.M. Abbas have argued that India's refusal to acknowledge the Ganges as an international river was the reason for the fruitless nature of the talks for these two decades. [lxxvii]

After 1971 however, things changed. Firstly, Bangladesh emerged as a newly independent nation, thereby creating new relationships in the South Asia region. This also changed the riparian structures and organization in the Ganges basin. Initially, the cooperation between Bangladesh and India reached a new high; there was a genuine sense of friendship and bilateral cooperation between the two neighbours, especially given India's role in the liberation of Bangladesh. [lxxviii] In 1972, both the Prime Ministers of India and Bangladesh agreed to establish a permanent Joint Rivers Commission (JRC), consisting of members from both nations. The commission's role was to draw up a wide-ranging survey of the river systems shared by the two countries; this was to contribute towards projects for flood control and water resources development for mutual benefits.[lxxix]

Unfortunately, bilateral cooperation over the Ganges Brahmaputra basin soon came to a standstill after it was discovered both India and Bangladesh had very different interests in utilizing and developing the Ganges water. The Farakka problem soon acquired major significance in bilateral relations and became a moot question in Indo-Bangladesh foreign policy. As a result the JRC found its hands tied; it too was caught between these two divergent interests.[lxxx]

In 1974 Sheikh Mujibur Rahman, the Bangladeshi Prime Minister and India's Prime Minister, Indira Gandhi issued a joint declaration agreeing to two basic principles on water sharing. First, there was a need to "augment" the Ganges water during the lean or dry season flows to meet the needs of both India and Bangladesh. Second, the augmentation was to take place through the "optimum utilization of the water resources available to both countries."[lxxxi] The JRC however, which took up the matter of augmentation, found it could not do much as both India and Bangladesh had completely different opinions on how to approach the problem.

In 1975, two things of importance happened: firstly, the Farakka barrage went on test operation for 41 days and for the next two years India would continue to draw water. Secondly, Sheikh Rahman was murdered and India was reluctant to negotiate with the new military regime. It was only in 1977 that Morarji Desai, India's then Prime Minister took an interest in resolving the Farakka problem; in November that year India and Bangladesh signed an agreement on water sharing of the Ganges River and augmenting its flows. [lxxxii]

The agreement was important for several reasons. It highlighted the importance of diplomacy and international politics in bilateral relations. Bangladesh had tried unsuccessfully to internationalize the problem. They had brought the matter up at the Islamic Foreign Ministers Conference in Istanbul in 1976 and at the 31st Session of the United Nations, the same year. Bangladesh asked for UN intervention on the grounds that the dispute had implications for both security as well as environmental repercussions in the region. India feeling the heat from the international community was keen to avoid further disagreement and asked

Bangladesh to return to the negotiating table. The agreement therefore marked its acceptance and return to bilateral relations over the matter. But the agreement was also heavily criticized in both nations. India was seen as having given away too much sovereignty to Bangladesh. Bangladesh in turn was criticized for the temporary nature of the agreement (it was for only 5 years). But aside from these criticisms there was a sense of relief in both nations and the agreement was welcomed as a solution to the problem of water sharing over the Ganges. [lxxxiii] However, the Farakka matter remained unresolved.

By the late 1980s after the second agreement had lapsed, India and Bangladesh were experiencing heightened tensions over the Farakka problem. Between 1988 and1996, both countries' positions over the matter were circumscribed by intense myopia. India felt Bangladesh was far too rigid, and had failed to appreciate the needs of upstream populations and had in turn, demonized India – whenever there were floods or drought in Bangladesh, it was blamed upon the Farakka Barrage. India believed Bangladesh was hugely overstating its water requirements and had done a great misdeed in turning the Farakka problem into one ridden with national or domestic political fault lines – this made any kind of cross border negotiation extremely difficult. [lxxxiv]

In 1996, a historic treaty was signed and Bangladesh and India finally reached a long-term and sustainable solution to share the waters of the Ganges.

The treaty itself is remarkable for several reasons. For the first time water sharing was delinked from augmentation. In addition, India formally recognized the rights of a lower riparian in sharing waters. Also important is the use of international treaties as a foundation for the agreement over the Ganges water. According to the Helsinki Rules on the Uses of the Waters of International Rivers, adopted by the International Law Association in 1966, "each basin state is entitled, within its territory, to a reasonable and equitable share in the beneficial uses of the waters of an international drainage basin." These rules, which Bangladesh considered India to be in violation of with regards to the Farakka Barrage, later became the partial basis of the Ganges Water Treaty of 1996.[lxxxv]

The 30-year timetable for the treaty has also provided stability in bilateral relations, a marked change from the short-term agreements of the past between the two countries. It serves as a template for sharing of water between India and other nations as well as for sharing water with Bangladesh in other river projects.

Policy Recommendations

While the dispute over the Ganges River may have been contentious and adversarial, it is this paper's contention that cooperation over management of the River and its resources for the Sundarbans will ultimately foster peace. This is evident in the stability achieved over the Ganges Treaty of 1996.

Need for an Environmental Impact Assessment

In the 1970s, no importance was given to an Environmental Impact Assessment (EIA) to measure the impact of the Farakka Barrage on the prevalent ecosystem.[lxxxvi] An EIA, together with a transnational impact or trans-boundary impact assessment ought to have been conducted. It is this report's recommendation that both these impact assessments be done immediately. Allocation of water in a basin cannot be determined simply on the basis of industrial, agricultural or human needs. Rivers contain highly complex aquatic ecosystems and the survival of coastal wetlands is entirely contingent on the equilibrium between fresh and salt water. Both India and Bangladesh must advocate the need for not only a bilateral, but also a regional policy, under the auspices of a body such as the South Asian Association for Regional Cooperation (SAARC) that focuses on an EIA for the entire Ganges-Brahmaputra Delta. The policy to implement an EIA must be accompanied with a provision to continue monitoring the Delta basin.

Augmentation is Not the Answer

This paper is against the augmentation proposals put forward by both India and Bangladesh to supplement water shortages in the lean season. Nor does this paper advocate augmentation to feed water shortages in the Sundarbans river system, especially in the South- Western areas. Instead, this paper proposes three alternative means of "water-resource management" in which both India and Bangladesh can work together to increase any perceived water shortages as a result of the Farakka Barrage construction.

There are three methods whereby India and Bangladesh can cooperate to devise as well as implement water-resource management policies.

Water-Harvesting

Focus on local water conservation methods such as constructing and rehabilitating ponds and lakes is a much better alternative to storage of water in large reservoirs in the Delta basin. Both India and Bangladesh can create appropriate decentralized institutions at the local level to oversee conservation and maintenance of these projects. Of particular interest is the revival of rainwater harvesting.

Through a bilateral agreement India and Bangladesh can work to develop [a] community-based approaches to harvesting water in the Sundarbans

Through a bilateral agreement India and Bangladesh can work to develop such community-based approaches to harvesting water in the Sundarbans and apply those very lessons for other water starved areas in their own nations. A regional framework for water harvesting will tackle problems such as riverbank erosion, soil conservation, arsenic contamination in groundwater and generation of hydropower. [lxxxvii]

Recycle Waste Water

Both India and Bangladesh possess large quantities of waste-water that goes untreated, thereby adding to their pollution woes. India and Bangladesh could reuse this waste- water for agriculture and aquaculture and lessen the demand on freshwater in the Sundarbans and in the region as a whole. Agreeing to a joint policy on water resource management would ensure access to each other's sources of waste- water for reuse.

Establish Correct Pricing Policies for Water

Approximately 80 percent of India's water is spent on irrigation, an investment which is seldom recovered.[lxxxviii]As a result, farmers have become wasteful in their usage of water and have a tendency to invest more in water-intensive crops.[lxxxix] In order to shore up more water for the Sundarbans river systems and ensure greater water supply to India and Bangladesh in general, India is beginning to create a more balanced tariff for extracting groundwater.[xc]India must revise irrigation and power prices for removal of groundwater.

In the case of Bangladesh, these prices need to be introduced. This is an area both India and Bangladesh can work together on, especially with regard to the Sundarbans. Creation of a joint tariff policy over the Sundarbans water resources and applicable to private sector industries as well as local communities would provide a solid base to export the same measures elsewhere in each country. The principles of pricing and market economics must be brought into any policy resolution between India and Bangladesh.

A Community Based Approach

Both India and Bangladesh need to invest in more community-based approaches to water conservation in the Sundarbans. The Centre for Ecology and Hydrology in Wallingford, England has developed a system of measurement called the "Water Poverty Index." This utilizes 5 means of measurement such as available resources; people's access to water; their capacity to afford and organize water supplies; use of water for domestic, industrial and environmental purposes; and finally, environmental management. [xci] Such an index allows for the creation of 'targeted' policies, where they need to be improved, or in some cases, made. There is scope for both India and Bangladesh to jointly create a similar index for the Sundarbans.

Using a localized, community based approach neighbouring villages could collect and compare the data, under the management of the border district administrations in both India and Bangladesh. Once this is done, a joint mechanism or body between India and Bangladesh could determine appropriate policies to mitigate environmental damage. For instance, an index of this nature will highlight where water salinity has increased; it will also localize the extent of human dependence on the water for both personal consumption as well as livelihood resource.

Improving Governance

There is an urgent need to tackle systemic problems such as lack of transparency and accountability. Both nations, through a joint task force can ensure that information is shared at all levels, from community, district, through to state levels, amongst all stakeholders by implementing systems to monitor hydrological data as well as to ensure it is consistently accurate. There are Water Resource Agencies in both nations, but only the one in Bangladesh actually carries out any tasks. [xcii]

No Inter-Linking of Rivers

India's proposal to inter-link its rivers (ILR) must be immediately scrapped. This ambitious project, the largest of its kind in the world, involves building several dams and thousands of miles of canals, a plan that will not only cause massive ecological damage and displacement of human settlements within India; it will also negatively impact bilateral relations with Bangladesh, the lower riparian. The projects aims to link 37 of India's major river by 2016 and 25 new dams are planned for the Ganges and Brahmaputra rivers alone, in the Himalayan River Basin. [xciii] Bangladesh is concerned that India's plan to divert the waters of the Brahmaputra to feed parched Indian territories will reduce the amount of water flowing from India to Bangladesh. The Brahmaputra, one of Asia's major rivers, has its source in South- Western Tibet. Nearly 3 thousand kilometres long, it winds its way through the Himalayas, and then through India to Bangladesh where it flows into the Bay of Bengal. [xciv]

The impacts of the ILR on Bangladesh will be the function of many variables, including the alteration of hydrology, river dynamics, ecosystem changes, agricultural productivity, intrusion of salinity and public health. [xcv] The Bangladeshi government has already taken up the problem with the Indian government and is urging them to reconsider the river-linking project. Using the Farakka Barrage as an example, Bangladesh can internationalize the problem, thereby increasing its bargaining power and share over the Ganges water. This will harm India's reputation in the global community, not to mention the already fragile relations with Bangladesh.

India however is not the only country with designs on diversion of the Brahmaputra's water. China has recently announced plans for a new anti-drought project, which involves diverting the Brahmaputra's waters to the parched Xinjiang region. This has raised serious concerns in both India and Bangladesh, although Beijing has assured the two countries there will be no downstream impact.

This is an opportunity for both India and Bangladesh to join hands and lead the way for further regional cooperation and effective monitoring of the waters of the Himalayan River Basin. A good place to start would be for India to cancel its ILR project, thereby preventing substantial ecological destruction in India and also reassuring Bangladesh it has its best interests in mind. Both nations should also work together to implement some of the recommendations of the Dhaka Declaration on Water Security; a statement issued in 2010, by over 25 distinguished water experts from India, Bangladesh, Nepal and China, part of a long term process to build confidence and collaboration between countries that make up the Himalayan River Basin. The declaration proposes an expert committee to prepare a road map for data-sharing and scientific exchange and to prepare guidelines for introducing transparency regarding relevant data. The declaration asks for "greater political commitment and data exchange among Himalaya basin countries for collective approaches to the region's water challenges".[xcvi]

CHAPTER THREE

The Maritime Boundaries Dispute and Climate Refugees

New Moore Island – or South Talpatti – sat in the Sundarbans mangrove delta in the mouth of the Hariabhanga River that divides India and Bangladesh. It no longer exists. But for close to thirty years, this small, uninhabited island in the Bay of Bengal was a source of contention between the two countries. That is, until it was submerged last year.[xcvii]

New Moore Island brought to notice one of the most pertinent problems in Indo-Bangladesh foreign policy: the maritime boundaries dispute.

History of the Maritime Boundaries Dispute

A major hurdle in determining maritime boundaries is the geographical position of both countries: they are both adjacent nations, not geographically opposite to each other. As a result, the method of equidistant delimitation does not apply in the case of India and Bangladesh. Bangladesh has asserted the need for equitable determination of maritime borders; India disagrees with this assessment. [xcviii]

In this context, there are several problems yet to be determined in delimiting maritime boundaries. First of course, is determining the river boundary of the Hariabhanga River in West Bengal, between India and Bangladesh. According to the Radcliffe Award (establishing the East Pakistan and India boundary in 1947), the 'mid-channel flow' principle or *thalweg* doctrine is generally recognized as the international boundary on river borders between these two countries. [xcix]

The remaining problems concern the 1982 UN Convention on the Law of the Sea (UNCLOS). Under this convention a state can claim 200 miles of its territorial waters of which the initial 12 miles are known as territorial waters and the remaining 188 as its Exclusive Economic Zone (EEZ). In some cases, owing to specific geo-physical characteristics, such as an extended seabed or continental shelf, a state can take this claim to 350 nautical miles. Neither India nor Bangladesh has ratified the demarcation of these boundaries. [c] But delimitation of the sea boundaries has over time become a hugely important problem. Both countries have mounting energy needs to offset demand from burgeoning populations. As a result, India and Bangladesh are both exploring their maritime zones for oil and gas, as well as allocating offshore blocks to foreign multinationals and overlapping claims of maritime zones have become a regular feature in the region. This has led to delays as each nation challenges the others' sovereignty in the Bay of Bengal. [ci]

Climate Refugees

Destruction of the Sundarbans due to rising sea levels and natural disasters could lead to 70,000 climate change refugees in the next 30 years. Already in 2009, there were an estimated 8,000 climate refugees in the Indian Sundarbans. [cii] The Sundarbans could become the epicentre for one of the biggest challenges to face human existence in the coming decades: the problem of "climate change refugees".

Experts at the School of Oceanographic Studies at Jadavpur University in 2003 completed a decade long study which claimed the sea is rising at 3.14 mm each year in the Sundarbans as against a world average of 2 mm. [ciii] According to a recently released UNESCO report, a sea level rise of 45 cm will result in destruction of more than 75 percent of the Sundarbans mangroves.[civ] A study carried out in 2001-2002 found that out of the existing 100 islands on the Indian side, 12 were susceptible to excessive erosion and would lose 15 percent of their landmass by 2015.[cv] Already in the last 30 years, erosion of the Indian Sundarbans island system has resulted in reduction of 100 km² of land.[cvi]Aside from the obvious devastation to the existing biodiversity this could also damage any possibility of sustaining current human habitation.

In the past five years in the Sundarbans, two-thirds of the Ghoramara islands are submerged while the Sagar islands have lost around 7,500 acres of land; it has seen seven villages disappear under the relentless tides. In the past two decades, four islands (Bedford, Lohachara, Kabasgadi and

Suparibhanga) have been permanently flooded and 6,000 families have been made homeless, many living in the slums of Dhaka. [cvii] And further north, Sagar Island already houses 20,000 refugees from the tides.

Cyclone Aila left over 2.3 million people displaced and nearly 8,000 missing between India and Bangladesh. [cviii] The Sundarbans itself was inundated with over 20 ft of water and an estimated 400,000 people there were marooned by flooding.

A 1 metre rise in sea level is projected to displace approximately 7.1 million people in India, and about 5,764 sq km of land area will be lost, along with 4,200 km of roads

There is more bad news. According to other reports the migratory climate models forecast that as the world warms, the rains will be concentrated in a shorter period, throwing up a devastating combination of severe floods and longer periods of drought. If the sea level rises by up to a metre this century, as many as 10 to 30 million Bangladeshis along the southern coast could become climate refugees. [cix]

A 1 metre rise in sea level is projected to displace approximately 7.1 million people in India, and about 5,764 sq km of land area will be lost, along with 4,200 km of roads.[cx]

Illegal Immigration across Borders

More than 5 million people in Bangladesh live in areas which are highly vulnerable to cyclones and storm surges. Globally, the most significant proportionate increase in populations exposed to the dangers of extreme climates will take place in two cities in Bangladesh: Dhaka and Chittagong. The problem of refugees has developed into one of the most pressing problems in Indo-Bangladesh relations. In 2001, the Union Home Minister said over 12 million illegal Bangladeshi immigrants are living in India, in 17 Indian states. However, these estimates were later withdrawn by the Home Ministry which said the data was both "unreliable" and "based on hearsay." The Bangladeshi government does not officially recognize those migrants; therefore, it does not provide help or support. In 2003, Bangladesh's foreign minister was quoted as saying that "not a single unauthorized Bangladeshi resided in India." [cxi]

Bangladesh currently faces a severe crisis of land and water, caused by population growth, environmental change and recurring natural disasters and the flow of migration from Bangladesh to India may increase at a faster rate. [cxii] The future effects of climate change are only likely to increase the flow of population from Bangladesh to India.

History of the Border Dispute

The border problem has plagued relations between the two countries ever since Bangladesh was carved out of Pakistan in 1971. The prime ministers of India and Bangladesh agreed during a state visit to New Delhi in January 2010 to stop illegal smuggling and deaths on the border. But India has gone a step further; in recent years it has built a barbed-wire fence along most of the 4,095-kilometer border at a cost of approximately Rs. 5,205 crore. [cxiii]

Thus far, India has 111 enclaves in Bangladeshi territory and Bangladesh has 51 enclaves inside India

Thus far, India has 111 enclaves in Bangladeshi territory and Bangladesh has 51 enclaves inside India. In addition, the fencing project was supposed to be constructed about 150 yards away from the hypothetical boundary line that separates Bangladesh and India. Unfortunately, in many cases this rule has also been flouted – often the fence has been constructed kilometres away from the boundary line and this "no-man's land" is inhabited by families. Almost 90,000 people in over 149 villages exist as though prisoners and the Indian state has abdicated all responsibility towards them.[cxiv] However, in July this year, both sides agreed to an historic first-ever joint census along the border. Both India and Bangladesh have also agreed to exchange enclaves in order to solve this long-standing territorial dispute and maintain that it is up to the people to determine where they want to live. [cxv]

However, earlier this year the Indian government agreed to equip its border security guards with non-lethal weapons to stop killing unarmed Bangladeshis along the border. [cxvi] Bangladesh has also in recent months been cooperating with India over the ULFA problem and has handed over a number of key fugitives to India. [cxvii]

Internationally Un-Recognized

The term "climate refugees" has, as yet, no place in international law. UNHCR, the United Nations refugee agency, does not recognize climate or environment refugees, as they are not listed under the UN's 1951 Refugee Convention. Now some experts suggest the Convention should be amended to allow for environmental displacement. [cxviii]

According to the UN's Intergovernmental Panel on Climate Change, there are around 25 million climate refugees which could increase to as many as 150 million by 2050. Another controversial study states that when global warming takes hold there could be as many as 200 million people displaced by disruptions of the monsoon system and other rainfall regimes, by droughts of unprecedented severity and duration, and by a rise in sea levels that will result in coastal flooding.[cxix] A major concern for India and Bangladesh is in determining responsibility for the protection and rehabilitation of these refugees, most of who currently reside in slums in Dhaka or Kolkata.

Policy Recommendations

Any steps taken by both India and Bangladesh in recognizing and rehabilitating climate refugees will serve as a paradigm for nations across the globe. This is a historic opportunity for both nations. Below are some suggestions.

Define "Climate Refugees"

There is a momentous opportunity here for both India and Bangladesh to work together and petition the international community to recognize climate refugees as a classification of refugees and to include both migration within the country and immigration between countries. This will guarantee individuals displaced by climate change natural rights under an international treaty.

Promote a Strong Micro-Insurance Programme for the Sundarbans

Both India and Bangladesh have traditionally very strong grassroots institutions that provide micro-credit to the individuals at the bottom of the pyramid. Managing the crises of climate change refugees could well be served by the use of such micro-credit techniques. Micro insurance is an innovative way to ensure lower risks for vulnerable households. It also becomes a registry or means of recording demographic populations of refugees on the basis of which identity cards can be issued. This will ensure both countries have a record of those with refugee status and it serves to lower illegal immigration. Both countries could set up micro-insurance for households in the Sundarbans, which are prone to natural disasters. The insurance can be disbursed through NGOs, Non-Banking Financial Corporations (NBFCs) or district agencies already familiar with handling micro credit in poor rural communities. The insurance should also cover any loss of livelihood.

Natural disaster insurance exists in the Maldives and it could serve as a good model from which both India and Bangladesh can learn valuable lessons.

Develop a Systemic Response

The Sundarbans are particularly prone to natural calamities and India has no proper system to deal with such calamities on this scale. A search of India's Natural Disaster Management Authority provided no reference of any adaptation programs for climate change, nor has India's National Action Plan on Climate Change (NAPCC) published in June 2008, provided any real initiatives on how adaptation to climate change is to be achieved. According to India's Prime Minister, Dr. Manmohan Singh, speaking at the G8 summit in July 2009, the country already spends 2-2.5% of its GDP on meeting the consequences of climatic conditions. Even the limit of a two-degree centigrade rise in average global temperature, agreed by the G8, would require a significant increase in public spending to meet its predicted environmental consequences.[cxx]

A National Task Force on Climate Change should mandate all relevant ministries to have nodal officials who are concerned with looking at climate change needs from the perspective of their ministry and be involved with planning and creating policies and action plans

The same situation persists in Bangladesh, with almost no national cohesive planning for prevention or management of natural disasters.

Rehabilitation of Climate Refugees

There is urgent need for a joint relocation and emergency evacuation program in the case of climate disasters such as cyclones or flooding - both of which are becoming annual events along the low-lying delta regions in Bangladesh and India.

A National Task Force on Climate Change should mandate all relevant ministries to have nodal officials who are concerned with looking at climate change needs from the perspective of their ministry and be involved with planning and creating policies and action plans. There should be a full study of the country's vulnerabilities and adaptation needs which includes analyzing impacts on water resources, agriculture, bio-diversity, ecosystems and human health. Anticipatory actions should be contemplated in all development programs and the act should ensure that the legal measures for both reactive and anticipatory adaptation needs are integrated into all relevant laws.

Restore the Capacity of the Mangroves

It is crucial that both India and Bangladesh undertake measures to increase the adaptive capacity of the Sundarbans mangroves against the adverse effects of sea-level rise. This would mean conservation of remaining mangrove forests in protected areas. Another idea would be to restore mangrove forests through re-planting selected mangrove tree species, for example along freshwater canals of reclaimed land; this has been successfully practiced on Sagar Island.[cxxicxxii]

In 1995, the United Nations Development Programme (UNDP) evaluated the cost of building 2,200 km of protective storm and flood embankments that could provide the same level of protection as the Sundarbans mangroves. The capital investment was estimated at about US $294 million with a yearly maintenance budget of US $6million. [cxxiii]

CHAPTER FOUR

Livelihood Resource Dependency and the Environment

The Sundarbans has a rapidly burgeoning population, (estimated at 2.5million in 1981, 3 million in 1991 and 4.5 million in 2001) rampant illiteracy, no modern transport or energy services, and no education or healthcare system.[cxxiv] On the Indian side the major source of revenue for the local economy is the Non-Timber Related Forest Produce (NTFP) and across both countries the main economic pursuits are agriculture, and any fishery related enterprises, such as shrimp cultivation. A recent study shows that the contribution of NTFPs is quite high as it contributes almost 79% (Rs. 80,000) on an average to the annual income of the collector's family [cxxv]

Increasing unemployment and poverty, rising sea levels, reduction in the mangrove cover and coastal erosion are having a punishing impact on land based livelihood activities

Increasing unemployment and poverty, rising sea levels, reduction in the mangrove cover and coastal erosion are having a punishing impact on land based livelihood activities. Consequently, the local populations have begun to exploit the natural resources of the Sundarbans in order to eke out basic subsistence, but are doing so in a manner that is simply unsustainable over the long term. [cxxvi] There is tremendous scope and opportunity for both India and Bangladesh to work together to ease the pressures local populations are placing on the environment, whilst providing the resources necessary for livelihood and development.

Livelihood Displacement

The loss of livelihood is a result of three major factors: climate conditions, inappropriate application of forest laws and unsustainable employment techniques.

Climatic conditions result in livelihood displacement as a result of submerged land, a problem related to climate refugees explained in the previous section. Loss of livelihood has also occurred due to inappropriate implementation of forest laws in both India and Bangladesh. These forest laws pertain to acquisition and ownership of property in areas designated as forest reserves, but over time they have resulted in loss of land for the inhabitants of the Sundarbans.

Loss of Land Due to Evolution of Forest Laws

India

The Indian Forest Act of 1927, modelled after the Forest Act of 1878, stipulates that all property belongs to the State. Under this law, all community rights were removed and the local inhabitants were completely alienated from forest management. cxxvii Local inhabitants had no rights to the forests under law and had no stake in forest conservation.

Prior to this, ownership of the forests rested with the princes and local chieftains but indigenous communities or inhabitants of the forests were allowed unrestricted access to the forest for their use. In 1807, the East India Company claimed royalty rights for teak and the power was bestowed on the Conservator of Forests for the management of the forests as a source of revenue. cxxviii After 1878 however, the forests came to be classified on three grounds: Reserve Forests, where the local community was completely alienated from forest usage; Protected Forests, where indigenous rights were recorded but not settled; and finally Village Forests, where local needs were to be met, but this remained true mostly on paper.

In 1878, the Sundarbans were declared partly Reserved and partly Protected Forest; the land was not open to conversion to agriculture without the consent of the forest department. The main problem with these classifications was that the areas, which were declared as reserved or protected, had indigenous people already living there. This made the tribal populations encroachers on their own land and liable to punishment. Their use of any forest-produce including timber and small twigs was banned. Agricultural practices in these areas were also banned.

The Scheduled Tribes and Other Traditional Forest Dwellers (Recognition of Forest Rights) Act, 2006 recognized the rights of forest-dwelling Scheduled Tribes and other traditional forest dwellers over the forest areas inhabited by them over generations. But there are lacunas even within this Act; because only land currently under cultivation can be claimed, it has led to a large scale loss of land without any compensation to the indigenous communities. [cxxix]

The resulting impact on biodiversity habitats such as the Sundarbans has been tremendous despite the fact that the Sundarbans have been made into a wildlife reserve where hunting and felling of trees is strictly prohibited. Excessive felling of trees for commercial purposes and for agriculture has led to the reduction of forest cover. Over the past 200 years, the Sundarbans have been reduced by 50 percent as a result of clearance for agriculture. There is also loss of wildlife due to poaching.[cxxx]

There have been many efforts by both government and international agencies to conserve Sundarbans' natural resources. A National Mangrove Committee comprised of experts and representatives of concerned government departments, was established in 1987. However, information regarding the forest cover and conditions remain hazy and public programs remain weak.[cxxxi] This year, the Ministry of Environment and Forests released a report on the Management effectiveness in the Sundarbans forests. The study, jointly prepared by National Tiger Conservation Authority (NTCA) and Wildlife Institute of India (WII), also said that inadequate inter-agency co-ordination, lack of proper research, unrestricted number of tourists and inadequate trained tourist guides are some of the problems that plague the Sundarbans forest reserve. [cxxxii]

Bangladesh

In Bangladesh, the contribution of the forestry sector to GDP is nearly 4 percent and the sector directly employs approximately 2.5 percent of the labour force. [cxxxiii] The Sundarbans were declared a Wildlife Sanctuary to conserve and protect animals in 1974 and over a decade later, the Sundarbans division of the Forest Department under the Ministry of Environment and Forests was created. In Bangladesh, as in India, the strictures of the Forest Act of 1878 have largely guided the current prevailing legal framework for Forests. Forest policy established under Pakistani rule showed a high degree of continuity with its colonial heritage and maintained an emphasis on commercial and industrial interests. This process of commercialization continued after the independence of Bangladesh and in 1979 the first national Forest Policy neglected the larger problem of broader stakeholder participation. [cxxxiv] However, an amendment was made in 2000 to introduce the concept of Community or Social Forestry, but this is considered poorly implemented. [cxxxv]I nstead, the rights of indigenous people and communities are inadequately recognized. There are misunderstandings between forest user groups and the Forest department and the declaration of Protected Areas is made without providing the affected group any alternative livelihood strategies. [cxxxvi]

In 1994 the National Forestry Policy was created as an amendment to the 1979 policy. The policy was formulated to initiate a 20 year Forestry Master Plan (FMP), with the assistance of the Asian Development Bank and the United Nations Development Program, to preserve and develop the nation's forest resources with a view to ensure sustainability, efficiency and people's participation.[cxxxvii]

Loss Due to Unsustainable Employment Techniques

In 2001, a socio-economic survey in Southkhali Union found that 49 percent of households directly extracted forest resources for their livelihood, including 98 percent of landless forest fishers and almost all were partially dependent on forest or natural resources. [cxxxviii] In India, a study conducted in select villages in the Sundarbans Tiger Reserve in 2010 found that nearly 79 percent of people in Sundarbans are engaged as agricultural labour. This agriculture, in spite of being the main occupation of the people in Sundarbans, is not high yielding due to salinity of the soil, which prevents optimum growth of agricultural crops. Around 50 percent of agricultural labourers are landless. Therefore, the Reserve Forest area serves as the buffer for their survival and though the percentages of NTFP collectors are between 6-9 percent, the contribution of NTFPs is 79 percent in the total annual household income. [cxxxix]

There is a clear scope for gentle regulation, in the interest of the very people who depend on the resources.

Shrimp Cultivation

No practice has had such a profoundly adverse impact on the biodiversity of the Sundarbans as has shrimp cultivation. There are currently two prevalent forms of shrimp cultivation in the Sundarbans; salt water and freshwater; both lead to increased salination of land.

Policy Recommendations

Abolish Cyclical Employment

Both India and Bangladesh should set up a joint program to share tourism revenue with locals. They should also coordinate vocational schemes for locals with an emphasis on efforts to save the Sunderbans. A focus on sustainable use of resources would be of great benefit in reducing the dependency of local populations on natural resources. In India alone, this region consists mainly of backward classes, which accounts for more than 45 percent of the total population as against the state figure of around 25 percent, who are dependent on various types of forest and non forest based NTFPs for their livelihood.[cxl] However, NTFP collection from Sundarbans should be levied on those who are commercially operating at a high scale of economy and those who collect in small quantities (mainly the rural poor) should be exempted from such taxations.

Use Micro-Credit and Micro-Insurance

Both India and Bangladesh are innovative leaders in providing credit to poor populations traditionally excluded from the financial system. Both governments can work together to invite micro-credit organizations to set up shop in the Sundarbans and provide a necessary and urgent means of financial support to the marginalized communities. Not only would this supplement employment revenue under state schemes such as the NREGA, it would provide the much-needed stimulus to foster entrepreneurial growth among poorer sections in the region.

Continue Emphasis on Stakeholder Participation

This paper strongly advocates a bilateral policy akin to Bangladesh's Nishogro Programme, established in 2004. This policy aims to create a forest department that is actively trying to manage land use near the Sundarbans through the Integrated Coastal Zone Management Project in Bangladesh. It also looks to influence poverty levels in the region through the creation of a formal institution, at the district administration level, along the forested Sundarbans border district areas, to bring together stakeholders to co-manage protected areas, while generating alternate incomes and advocating for better management policies. It also seeks to develop the institutional capacity of the Forest Department and develop infrastructure in protected areas as well as restore them.

A good example to follow is that of the International Gorilla Conservation Programme (IGCP) in the Virunga-Bwindi region in Central Africa since 1991 discussed in greater detail in the Conclusions section.

Promote Aquaculture

Both India and Bangladesh need to promote environmentally friendly and socially responsive shrimp farming by introducing internationally accepted quality control measures under a bilateral agreement, such as those stipulated under the FAO Code of Conduct for Responsible Fisheries (which also applies to aquaculture). Also, crab culture, pearl culture, and sea grass should be promoted over shrimp farming. India and Bangladesh should provide grassroots workshops, organized by district or zonal officers, to educate local populations of the harmful impact of unsustainable agricultural farming techniques. A good example here would be that of Thailand. As of 2006, it produced nearly 500,000 tonnes per year, 85 percent of which was exported to the US, Japan, EU and others. In addition, 85 percent of its shrimp farmers are small scale farmers having typically less than 3 hectares and often less than 1 hectare of land; these were family owned and run operations, just like India.[cxli]Thailand has had tremendous success in building an environmentally sustainable industry that corresponds to international quality standards with robust labour laws. For instance, the Department of Fisheries in Thailand has encouraged the creation of farmer groups through which it ensures standardization of farming practices, market access and internal auditing systems.[cxlii]

Promote Ecotourism

The Sundarbans is gradually becoming a destination for Ecotourism, camps for which have been set up in several areas. However, the involvement of local communities in ecotourism is very limited; there remains much scope for the government to involve local poor in ecotourism.[cxliii] This has yielded marked results in certain areas of the world where it has been tried.

Conclusion

As mentioned in the Introduction, there are several examples from around the world of trans-boundary environmental agreements that demonstrate how different countries have cooperated on environmental problems. Such an effort has helped in dealing with the specific concerns at hand and has also removed a sore spot in bilateral relations. Below we give some examples.

Rwanda, Uganda and the Democratic Republic of Congo (DRC)

The International Gorilla Conservation Programme (IGCP)[cxliv] is an excellent example of trans-boundary management of natural resources among three nations and multiple international agencies. The IGCP has been working in the Virunga-Bwindi region in Central Africa since 1991. The program is a coalition of the African Wildlife Foundation (AWF), Fauna and Flora International (FFI), and World Wide Fund for Nature (WWF). IGCP's mission is the conservation of mountain gorillas and the forests in Rwanda, the Democratic Republic of Congo and Uganda. These forests are spread out along the borders of these countries and are separated into four national parks. Prior to the arrival of IGCP, the four parks were managed as separate entities by the national protected area authorities.

High population density, human encroachment, poaching, deforestation and civil unrest all threaten the forest habitats. Pertinent to India and Bangladesh are the terms that dictate each country's manner of participation in and responsibilities toward the project. For instance, the governments of the three countries have given the IGCP the mandate to develop a regional framework and mechanisms for collaboration. In addition, it was independently agreed by the responsible ministries in each country that the three protected area authorities would participate as national representatives and form an integral part of the IGCP team. The authorities work as a team to manage the forest blocks as shared units and thereby strengthen conservation. A good illustration of this is a common communications protocol including radio links between the park headquarters. There are also quarterly regional meetings, bringing together key protected area authority staff from the three countries. The key emphasis here for regional cooperation is that of constant and regular communication.

U.S. and Mexico

The Border Environment Cooperation Commission (BECC) was established in 1993 under the Free Trade Treaty.[cxlv] It seeks to identify and coordinate environmental infrastructure projects on the borders of the two countries with specific reference to drinking water and waste management concerns. To date, the BECC reports significant progress in the Mexico-United States border region, with 19 projects on both sides of the border at an estimated cost of US$340 million benefiting some 6.4 million people. The project's uniqueness lies in the fact that it is largely driven on community participation. The commission's most significant achievement is the public process involved in both the definition and implementation of rules and procedures and the provision of information which is done right in the midst of the communities where projects are being considered. This is a key learning for any bilateral agreement between India and Bangladesh: the public process allows the commission a close and direct relationship with the communities, which are guaranteed a voice and ascertain that their infrastructure needs are being met. Not only does social validation thereby guarantee majority approval for the project, it ensures citizens access to information.

The Involvement of West Bengal

In addition to drawing from the experiences of nations that have cooperated to solve environmental problems, we would like to mention that West Bengal can play a crucial role. The state government of West Bengal must be directly involved in the management of the Sundarbans under a bilateral agreement if this idea is to succeed. Suggested areas of involvement are:

1. West Bengal has been largely successful in the implementation of a Wildlife Insurance scheme to tribal communities that live in the buffer zones of the Sundarbans.[cxlvi] This scheme could be expanded to include micro credit and insurance against disasters for the same communities. The West Bengal government can take a lead in implementation of such schemes, through its district administrations, under a bilateral agreement with Bangladesh over the Sundarbans ecosystem.

2. West Bengal Forest Department (FD) overseen by the Chief Conservator of Forests can play an instrumental role in setting up village development funds which are co-managed by the village communities. The FD could receive a portion of the funds under the bilateral agreement to portion out to these communities for creation of community assets. The FD can also use the funds to provide credit to individuals or groups for income generating activities.

3. A portion of the revenues generated from ecotourism in the Sundarbans could be used by the West Bengal government through the FD, and funnelled into conservation and community development activities. Locals should also be involved in eco-tourism and the revenues should be shared between both countries. This provides incentives for locals to conserve the Sunderbans as well.

4. West Bengal can also play a key role in scientific research on mangrove protection and biosphere management. The state government should be allocated funds under the bilateral agreement to set up a national institute for mangroves and coastal protection. The state government must also build capacity at Calcutta University and Institute of Environmental Studies and Wetland Management so that they can better serve the problems facing the Sundarbans today.

5. West Bengal's Forest Department, the state government and district officers can be instrumental in the conservation of the endangered Bengal tiger. Working with Bangladesh, the state government should set up a Tiger Conservation Foundation for the Sundarbans, which could be empowered to implement cross-border tiger population surveys and conservation methodologies through effective district management in both countries.

Bibliography

UNEP. "Freshwater Under Threat: South Asia." UNEP, Nairobi, 2008, 29.

UNESCO. " IUCN Evaluation Reports for the World Heritage Site." UNESCO, 1997.

UNESCO. *Case Studies on Climate Change and World Heritage.* UNESCO World Heritage Centre, 2007.

UNESCO. "IUCN Evaluation Reports for the World Heritage Sites." UNESCO, 1987.

UNDP, FAO, Government of Bangladesh. "Integrated Resource Development of the Sundarbans Reserved Forest." Draft Report, 1995.

National Communication (NATCOM). ""Chapter 3: Vulnerability Assessment and Adaptation"." India's Initial National Communication to the UNFCCC, Ministry of Environment and Forests, Government of India, 2004.

Sharma, Ram, Philip DeCosse, Monoj Roy, Mamun Khan, and Azhar Mazumder. "Co-Management of Protected Areas in South Asia with special reference to Bangladesh." Nishorgo Support Project , Nishorgo , Dhaka.

Strategic Foresight Group. "The Himalayan Challenge: Water Security in Emerging Asia, 2010." Strategic Foresight Group, 2010.

WWF. "Sundarbans: Future Imperfect Climate Adaptation Report." 2010, 9.

Waterbury, John. *Hydropolitics of the Nile Valley.* Syracuse University Press, 1979.

West Bengal Pollution Control Board. "Water Resources and its Quality in West Bengal, Status of Environment Report." West Bengal Pollution Control Board, 2009.

VOA News. *Bangladesh Denies Any Nationals Living Illegally In India.* January 8, 2003. http://www.voanews.com/bangla/news/a-16-a-2003-01-08-1-Bangladesh-94339969.html (accessed July 25, 2011).

Ahmad, Maimuna. *Farakka Barrage: The river is sometimes dryer on the other side.* http://www.mtholyoke.edu/~ahmad20m/politics/law.html (accessed May 2011).

Ahmad, QK, and AU Ahmed. "Regional cooperation in flood management in the Ganges-Brahmaputra-Meghna Region: Bangladesh perspective." *Natural Hazards* 28 (2003): 181-198.

Ahmad, QK, BG Verghese, RR Iyer, BB Pradhan, and SK Malla. *Converting Water into Wealth: Regional Cooperation in Harnessing the Eastern Himalayan Rivers.* Academic Publishers, 1994.

Ahmed, A T Salahuddin. "Dangers from India's interlinking of rivers project." *Holiday*, June 25, 2010.

Alam, S. ""Environmentally Induced Migration from Bangladesh to India"." *Strategic Analysis* 27, no. 3 (2003).

Alamgir, Jalal, and Bina D'Costa. "The 1971 Genocide: War Crimes and Political Crimes." *Economic and Political Weekly* 46, no. 13 (March 2011): 38-41.

Anam, Tahmima. "How Bangladeshis see India." *The Guardian*, August 14, 2007.

Anbarasan, Ethirajan. "Bangladesh and India begin joint census of border areas." *BBC News, Dhaka.* BBC News. July 14, 2011. http://www.bbc.co.uk/news/world-south-asia-14149042 (accessed July 25, 2011).

AsiaNews.it. "Chinese dams on the Brahmaputra threaten lives of Indians and Bangladeshis ." *AsiaNews.it.* June 18, 2011. http://www.asianews.it/news-en/Chinese-dams-on-the-Brahmaputra-threaten-lives-of-Indians-and-Bangladeshis-21869.html (accessed August 3, 2011).

Association, The International Law. *The Helsinki Rules on the Uses of the Waters of International River.* Committee on the Uses of the Waters of International Rivers, The International Law Association, Helsinki: UNESCO, 1967, 1-5.

Arsenic Crisis Information Centre. "Arsenic in the Main Report, Draft Development Strategy, National Water Management Plan." Arsenic Crisis Information Centre, Bangladesh, 2001, 4.

Banerjee, Bidisha. "The Great Wall of India." *Slate*, December 20, 2010.

Banerjee, Manisha. "A REPORT ON THE IMPACT OF FARAKKA BARRAGE ON THE HUMAN FABRIC." South Asian Network on Dams, Rivers and People (SANDRP), Kolkata, 1999.

Bandyopadhyay, Somnath, H B Soumya, and Parth J Shah. "Briefing Paper on Forest Policy, Community Stewardship and Management." Centre for Civil Society, 2005, 11.

Bangladesh Bureau of Statistics. "Statistical Year Book of Bangladesh." Ministry of Planning, Government of the People's Republic of Bangladesh, Dhaka, 2000.

Belt, Don. "The Coming Storm." *The National Geographic*, May 2011.

Bhardwaj, Sanjay. "Bangladesh Foreign Policy vis-a-vis India." *Strategic Analysis* (The Institute for Defence Studies and Analyses) 27, no. 2 (June 2003): 263-278.

Bhattacharya, AK, S Jha, and A Dave. "Biodiversity Conservation and Ecotourism; Lessons from Sundarban Tiger Reserve." *Van Vigyan-Jour For Science* 39 (2001): 73-81.

Border Environment Cooperation Commission. 2009. http://www.cocef.org/english/index.html (accessed July 7, 2011).

Brown, O. ""Climate Change and Forced Migration: Observations, Projections and Implications"." Background paper for the 2007 Human Development Report., 2007.

Centre for Science and Environment. *It isnt Agriculture.* http://www.cseindia.org/dte-supplement/industry20040215/agriculture.htm.

Chaudhuri, AB, and A Choudhury. "Mangroves of the Sundarbans: Volume One." Edited by G Acharya M Zakir Hussain. *IUCN- The World Conservation Union.* 1994.

Chakravarty, S. R. *Bangladesh Under Mujib Zia and Ershad: Dilemma of a New Nation.* Har-Anand Publications , 1995.

Chatterjee, Mohua. "Mamata likely to accompany PM in Bangladesh trip in Sept ." *The Times of India*, July 9, 2011.

Chowdhury, Md. Tamimul Alam. *Resource-dependent livelihoods in the Sundarbans.* Center for River Basin Organizations and Management, CRBOM Small Publications, 2010, 1-7.

Chowdhury, Md. Tamimul Alam. "Resource-dependent livelihoods in the Sundarbans." Center for River Basin Organizations and Management, Java, 2010.

Danda, Anamitra Anurag. "Surviving in the Sundarbans: Threats and Responses." University of Twente, 2007, 199.

Department of Fisheries, Thailand. "Thailand Experience and Opportunities for Aquaculture Certification." Department of Fisheries, Thailand Government, 2006.

Dixit, J.N. *My South Block years: memoirs of a foreign secretary.* UBS Publishers' Distributors Ltd, 1997.

Dhar, Sujoy. "Bangladesh's young smugglers risk their lives." *AlJazeera*, June 29, 2011.

—. "Hungry tides in India's Sundarbans." *AlJazeera*, December 14, 2009.

—. "Rising Seas Threaten Bengal's Deltaic People." *Inter Press Service*, June 4, 2007.

—. "Millions displaced by cyclone in India, Bangladesh." *Reuters*. Reuters. May 27, 2009. http://www.reuters.com/article/2009/05/27/us-cyclone-india-bangladesh-idUSTRE54Q27620090527 (accessed May 27, 2011).

Drew, Patrick, and Archbishop Desmond Tutu. *100 Places to Go Before They Disappear*. Abrams, 2011.

Guha, Ramachandra. 1998: 95.

Gupta, Alok Kumar. "Indo-Bangladesh Maritime Border Dispute: Problems and Prospects." *Institute of Peace and Conflict Studies*. October 7, 2008. http://www.ipcs.org/article/india/indo-bangladesh-maritime-border-dispute-problems-and-prospects-2699.html (accessed July 2, 2011).

Governement of India. "Census of India 2001." New Delhi, 2001.

Gopalganje Sundarban Global Integrated Rural Welfare Society . *www.savesundarbans.org/aila*. www.savesundarbans.org (accessed July 25, 2011).

Iyer, Ramaswamy R. "India's Water Relations with her Neighbours." The India-China Institute,, The New School University, New York, 2008.

IANS. "Sunderbans needs synchronised disaster management policy: Experts." *IANS*, July 15, 2009.

India Today. "Indo-Bangla Accord: Defying the Current." January 1997, 1997: 98-99.

Indian Express. "BJP Unhappy with Ganga Waters Pact." December 1996, 1996.

International Gorilla Conservation Programme. *International Gorilla Conservation Programme*. 2011. http://www.igcp.org/ (accessed July 25, 2011).

International Organization for Migration. *Migration and Climate Change: IOM Migration Research Series*. Geneva: International Organization for Migration, 2008.

Islam, Shafi, and Albrecht Gnauck. "Mangrove wetland ecosystems in Ganges-Brahmaputra delta in Bangladesh." *Frontiers of Earth Science in China* 2, no. 4 (December 2008): 439-448.

Human Rights Watch. ""Trigger Happy" Excessive Use of Force by Indian Troops at the Bangladesh Border." Human Rights Watch, 2010.

Huq, Saleemul, Syed Iqbal Ali, and A. Atiq Rahman. "Sea-Level Rise and Bangladesh: A Preliminary Analysis." *Journal of Coastal Research* (Coastal Education & Research Foundation, Inc.) Special Issue, no. 14 (1995): 44-53.

Hazra, Dr. Sugata. "Sundarban: Climate Change Adaptation and Mitigation Efforts." School of Oceanographic Studies, Jadavpur University, Kolkata.

Habib, Haroon. "Entente across the border." *The Hindu*, August 2, 2011.

Holiday. "Water Treaty Heading for Waterless Grave?" *Holiday*, March 14, 1997.

Homer-Dixon, Thomas. ""Environmental Scarcities and Violent Conflict: Evidence from Cases"." *International Security* 19, no. 1 (1994).

Hoq, M. Enamul. " An analysis of fisheries exploitation and management practices in Sundarbans mangrove ecosystem, Bangladesh." *Ocean & Coastal Management* 50 (January 2007): 411-427.

Hossain, Ishtaq. "Bangladesh-India Relations: The Ganges Water-Sharing Treaty and Beyond." *Asian Affairs* (Taylor & Francis, Ltd) 25, no. 3 (Fall 1998): 131-150.

Jai Jai Din. *Jai Jai Din*, December 24, 1996: 8-9.

Jeevan, S.S. "Orphans of the River." *Science and Environment Fortnightly Down to Earth* (Centre for Science and Environment), Februrary 2002: 28-37.

Kumar, Dr. Anand. "'Oil Poaching' Controversy in Bay of Bengal." 1877, South Asia Analysis Group, 2006.

Kashyap, Samudra Gupta. "Bangladesh to hand over top ULFA man Anup Chetia." *Indian Express*, August 4, 2011.

Karlekar, Hiranmay. "Hasina's visit and after." *The Daily Pioneer*, January 16, 2010.

Keane, David. "'The Environmental Causes and Consequences of Migration: A Search for the Meaning of 'Environmental Refugees'." *Georgetown International Environmental Law Review* 16 (2004): 209.

Kiran, Raj. "India aims to deport 20 million Bangladeshis." *Independent Media Centre India.* January 9, 2003. http://www.india.indymedia.org/en/2003/01/2730.shtml (accessed July 25, 2011).

Naujoks, Daniel. *Emigration, Immigration, and Diaspora Relations in India.* Migration Policy Institute. October 2009. http://www.migrationinformation.org/Profiles/display.cfm?ID=745#16 (accessed May 27, 2011).

Nautilus Consultants Ltd. "Investment Mechanisms for Socially and Environmentally Responsible Shrimp Culture." Nautilus Consultants Ltd and IIED, 2003, 160.

National Tiger Conservation Authority (NTCA) and Wildlife Institute of India (WII). "Management Effectiveness Evaluation (MEE) of Tiger Reserves." Ministry of Environment and Forests, Government of India, 2011.

Nishorgo Network. *Nishorgo Network.* 2010. http://www.nishorgo.org/nishorgo2/network.php (accessed July 25, 2011).

Mukharji, Deb. "A return to democracy." *www.india-seminar.com.* 2008. http://www.india-seminar.com/2008/584/584_deb_mukharji.htm (accessed July 28, 2011).

Mukherjee, Krishnendu. "Centre's lens on Sunderbans." *Times of India*, July 30, 2011.

Myers, N. ""Environmental Refugees: A Growing Phenomenon of the 21st Century",." *Philosophical Transactions of the Royal Society* 357 (2002): 609-13.

Myers, N. "Environmental Refugees." *Population and Environment* 19, no. 2 (1997): 1678.

Myers, N. "Environmental Refugees." *Population and Environment* 19, no. 2 (1997): 167-82.

Mehrotra-Khanna, Dr. Mansi. "Security Challenges to India-Bangladesh Relations." Centre for Land Welfare Studies, Delhi, 2010, 75.

Milam, William B. *Bangladesh and Pakistan Flirting with failure in South Asia .* Hurst, 2009.

Millat-e-Mustafa. "A view on Bangladesh Forest Policy trends in Bangladesh." Institute of Forestry and Environmental Sciences, University of Chittagong, 2002.

Ministry of Information and Broadcasting. "Bangladesh Progress." Department of Publications, Government of The People's Republic of Bangladesh, 1972.

Mookherjee, Nayanika. "Love in the Time of 1971: The Furore over Meherjaan." *Economic and Political Weekly* 46, no. 12 (March 2011): 25-27.

OECD. *Bridge Over Troubled Waters: Linking Climate Change and Development.* OECD, OECD Publishing, 2005, 153.

Panda, Architesh. "Climate Refugees: Implications for India." *Economic and Political Weekly* 45, no. 20 (May 2010): 76-79.

Potkin, Alan. "Watering the Bangladeshi Sundarbans." In *The Ganges water diversion: environmental effects and implications* , by M. Monirul Qader Mirza, 163-176. Kluwer Academic Publishers, 2004.

Prudhomme, Christel. "Assessing the Water Poverty Index in a Context of Climatic Changes." *http://www.nerc-wallingford.ac.uk.* CEH Wallingford. http://www.nerc-wallingford.ac.uk/ih/www/research/WPI/Appendices%209.13to9.14.pdf (accessed July 25, 2011).

Prakash, Anoop, and Shailaja Menon. "Fenced Indians Pay for 'Security'." *Economic and Political Weekly*, March 19, 2011: 33-37.

Swain, Ashok. "Displacing the Conflict: Environmental Destruction in Bangladesh and Ethnic Conflict in India." *Journal of Peace Research* (Sage Publications Ltd.) 33, no. 2 (May 1996): 189-20.

Saha, S, and A Choudhury. "Vegetation Analysis of Restored and Natural Mangrove Forest in Sagar Island, Sundarbans, East Coast of India." *Indian Journal of Marine Science* 24, no. 3 (1995): 133-136.

Sanyal, P. "Sea-Level Rise and Sundarban Mangrove, in G. Quadros, (ed)." *Proceedings of the National Seminar on Creeks, Estuaries and Mangroves – Pollution and Conservation.* Thane, 2002. 47-50.

Samad, Saleem. "India Vows to Rein In Trigger Happy Soldiers Along Bangladesh Border." *AHN*, March 15, 2011.

Samarakoon, Jayampathy. "Issues of Livelihood, Sustainable Development, and Governance: Bay of Bengal." *Ambio* (Royal Swedish Academy of Sciences) 33, no. 1/2 (Februrary 2004).

Sarkar, Sonia. "Running with the wolves." *The Telegraph, India*, July 24, 2011.

Sen, Soham G. "Conservation of the Sundarbans in Bangladesh through Sustainable Shrimp Aquaculture." Harvard Kennedy School, 2010, 55.

Singh, Anshu, Prodyut Bhattacharya, Pradeep Vyas, and Sarvashish Roy. "Contribution of NTFPs in the Livelihood of Mangrove Forest Dwellers of Sundarban." *Journal of Human Ecology* 29, no. 3 (2010): 191-200.

Stebbing, EP. *The Forests of India Vol. I.* 1923.

Strategic Foresight Group (SFG) and Bangladesh Institute of Peace and Security Studies (BIPSS) . "Dhaka Declaration on Water Security." *Second International Workshop on Himalayan Sub-regional Cooperation for Water Security.* Dhaka, 2010.

Ravi, N. "India doing a great deal on climate change: Manmohan." *The Hindu*, July 11, 2009.

Rahman , M Mizanur , and Javed Ameer. "Bring disaster support under legal framework ." *The Financial Express*, July 6, 2011.

Rangachari, R, and BG Verghese. "Making water work to translate poverty into prosperity: The Ganga-Brahmaputra-Barak region." In *Ganges-Brahmaputra-Meghna Region: A Framework for Sustainable Development*, by QK Ahmad, Asit K Biswas, R Rangachari and MM Sainju, 81-142. The University Press Kimited, 2001.

Rashid, Harun ur. "Bangladesh-India Maritime Boundary." *Institute of Peace and Conflict Studies.* February 1 11, 2009. http://www.ipcs.org/article/india-the-world/bangladesh-india-maritime-boundary-2805.html.

Rashid, Harun Ur. "35th anniversary of Bangladesh-China diplomatic ties." *The Daily Star*, October 20, 2010.

Rivers for Life. "Interlinking of Rivers – An overview." 16.

Roy, Bhaskar. "Can India and Bangladesh Create a Win-Win?" No.447, Chennai Centre for China Studies, Chennai, 2010.

Roy, Pinaki. "Climate refugees of the future." *International Institute for Environment and Development.* March 31, 2009. http://www.iied.org/climate-change/key-issues/community-based-adaptation/climate-refugees-future (accessed July 25, 2011).

Tapas, Paul. "Implementation Status & Results India Integrated Coastal Zone Management (P097985)." Projects and Operations: Integrated Coastal Zone Management, The World Bank, 2011, 5.

Tiwary, Rakesh. "Conflicts over International Waters." *Economic and Political Weekly* 41, no. 17 (April 2006): 1684-1692.

Times of India. "New Moore isle no more, expert blames warming." *Times of India*, March 25, 2010.

The Economist. "Priceless." *The Economist*, July 17, 2003.

The Hindu. "Accord on Ganga Hailed." *The Hindu*, December 21, 1996.

The Law Commission of India. "The Proposed Foreigners (Amendment) Bill, 2000." The Law Commission of India, 2000.

The New Nation. "Ganges Water Treaty Coming Apart." *The New Nation*, April 13, 1997.

The Telegraph. "Sea Change." *The Telegraph.* June 14, 2009. http://www.telegraphindia.com/1090614/jsp/calcutta/story_11105736.jsp (accessed May 27, 2011).

Citations

[i](Chatterjee 2011)
[ii](Rashid 2010)
[iii](Karlekar 2010)
[iv](B. Banerjee 2010)
[v](Panda 2010)
[vi](N. Myers 1997)
[vii](International Organization for Migration 2008)
[viii](Roy 2009)
[ix](IANS 2009)
[x](Gopalganje Sundarban Global Integrated Rural Welfare Society n.d.)
[xi](Tapas 2011)
[xii](Rahman and Ameer 2011)
[xiii](Sen 2010)
[xiv](The Law Commission of India 2000)
[xv](Sarkar 2011)
[xvi](Dhar, Bangladesh's young smugglers risk their lives 2011)
[xvii](Mukherjee 2011)
[xviii](Danda 2007)
[xix](Sen 2010)
[xx](Governement of India 2001)
[xxi](Hoq 2007)
[xxii](Chowdhury, Resource-dependent livelihoods in the Sundarbans 2010)
[xxiii](Nishorgo Network 2010)
[xxiv](Milam 2009)
[xxv](Mookherjee 2011)
[xxvi](Alamgir and D'Costa 2011)
[xxvii](Ministry of Information and Broadcasting 1972)
[xxviii](Bhardwaj 2003)
[xxix](Anam 2007)
[xxx](B. Roy 2010)

[xxxi](Bhardwaj 2003)
[xxxii](Mukharji 2008)
[xxxiii](Habib 2011)
[xxxiv](Anam 2007)
[xxxv](Bhardwaj 2003)
[xxxvi](Dixit 1997)
[xxxvii](Bhardwaj 2003)
[xxxviii](Dixit 1997)
[xxxix](Anam 2007)
[xl](Mukharji 2008)
[xli](Chakravarty 1995)
[xlii](B. Roy 2010)

[xliii](Rangachari and Verghese 2001)
[xliv](Samarakoon 2004)
[xlv](Ahmad, et al. 1994)
[xlvi](Samarakoon 2004)
[xlvii](UNEP 2008)
[xlviii](UNEP 2008)

[xlix](Samarakoon 2004)
[l](Oxfam, Oxfam Bangladesh 2011)
[li](OECD 2005)
[lii](UNESCO 1987)
[liii](UNESCO 1997)
[liv](Strategic Foresight Group 2010)
[lv](Potkin 2004)
[lvi](Islam and Gnauck 2008)
[lvii](Islam and Gnauck 2008)
[lviii](Islam and Gnauck 2008)
[lix](Samarakoon 2004)
[lx](Swain 1996)
[lxi](Banerjee 1999)
[lxii](West Bengal Pollution Control Board 2009)
[lxiii](Jeevan 2002)
[lxiv](Banerjee 1999)
[lxv](Samarakoon 2004)
[lxvi](Arsenic Crisis Information Centre 2001)
[lxvii](Huq, Ali and Rahman 1995)
[lxviii](Hazra n.d.)
[lxix](Waterbury 1979)
[lxx](Hossain 1998)
[lxxi](The Hindu 1996) and (Indian Express 1996)
[lxxii] (Indo-Bangla Accord: Defying the Current 1997)
[lxxiii](Jai Jai Din 1996)
[lxxiv](Holiday 1997) and (The New Nation 1997)
[lxxv](Hossain 1998)
[lxxvi](Tiwary 2006)
[lxxvii](Tiwary 2006)
[lxxviii](Tiwary 2006)
[lxxix](Tiwary 2006)
[lxxx](Tiwary 2006)
[lxxxi](Tiwary 2006)
[lxxxii](Tiwary 2006)
[lxxxiii](Tiwary 2006)
[lxxxiv](Iyer 2008)
[lxxxv](Association 1967)
[lxxxvi](Samarakoon 2004)
[lxxxvii](Ahmad and Ahmed 2003)
[lxxxviii](Centre for Science and Environment n.d.)
[lxxxix](The Economist 2003)
[xc](Samarakoon 2004)
[xci](Prudhomme n.d.)
[xcii](Samarakoon 2004)
[xciii](Rivers for Life n.d.)
[xciv](AsiaNews.it 2011)
[xcv](Ahmed 2010)
[xcvi](Strategic Foresight Group (SFG) and Bangladesh Institute of Peace and Security Studies (BIPSS) 2010)
[xcvii](Times of India 2010)
[xcviii](Gupta 2008)
[xcix](H. u. Rashid 2009)

c(H. u. Rashid 2009)
ci(Kumar 2006)
cii (Dhar, Hungry tides in India's Sundarbans 2009)
ciii(Dhar, Rising Seas Threaten Bengal's Deltaic People 2007)
civ(UNESCO 2007)
cv(WWF 2010)
cvi(Hazra n.d.)
cvii(WWF 2010)
cviii(Dhar 2009)
cix(Belt 2011)
cx(National Communication (NATCOM) 2004)
cxi(VOA News 2003)
cxii(Alam 2003)
cxiii(Prakash and Menon 2011)
cxiv(Prakash and Menon 2011)
cxv(Anbarasan 2011)
cxvi(Samad 2011)
cxvii(Kashyap 2011)
cxviii(Panda 2010)
cxix(Myers, "Environmental Refugees" 1997)
cxx(Ravi 2009)
cxxi(Saha and Choudhury 1995)
cxxii(Sanyal 2002)
cxxiii(UNDP, FAO, Government of Bangladesh 1995)
cxxiv(Danda 2007)
cxxv(Singh, et al. 2010)
cxxvi(Singh, et al. 2010)
cxxvii(Guha 1998)
cxxviii(Stebbing 1923)
cxxix(Bandyopadhyay, Soumya and Shah 2005)
cxxx(Chowdhury, Resource-dependent livelihoods in the Sundarbans 2010)
cxxxi(Chaudhuri and Choudhury 1994)
cxxxii(National Tiger Conservation Authority (NTCA) and Wildlife Institute of India (WII) 2011)
cxxxiii(Bangladesh Bureau of Statistics 2000)
cxxxiv(Millat-e-Mustafa 2002)
cxxxv(Millat-e-Mustafa 2002)
cxxxvi(Millat-e-Mustafa 2002)
cxxxvii(Millat-e-Mustafa 2002)
cxxxviii(Chowdhury 2010)
cxxxix(Singh, et al. 2010)
cxl(Singh, et al. 2010)
cxli(Nautilus Consultants Ltd 2003)
cxlii(Department of Fisheries, Thailand 2006)
cxliii(Bhattacharya, Jha and Dave 2001)
cxliv(International Gorilla Conservation Programme 2011)
cxlv(Border Environment Cooperation Commission 2009)
cxlvi(Sharma, et al. n.d.)